Learn About

ANIMALS

UNEXPECTED FLYERS

BY JAY LESLIE

Children's Press®
An imprint of Scholastic Inc.

A special thank-you to the Cincinnati Zoo & Botanical Garden for their expert consultation.

Library of Congress Cataloging-in-Publication Data available

ISBN 978-1-5461-0125-3 (library binding) | ISBN 978-1-5461-0126-0 (paperback)

10 9 8 7 6 5 4 3 2 1 25 26 27 28 29

Printed in China 62

First edition, 2025

Book design by Kay Petronio

Photos ©: cover, 1: Quentin Martinez/Biosphoto; 5 top right: Lea Scaddan/Getty Images; 7 main: Cede Prudente/Getty Images; 8–9: Cede Prudente/Getty Images; 10 top: Scott Linstead/Science Source; 11 main: Chien Lee/Minden Pictures; 14 top: Andrey Nekrasov/Getty Images; 15 main: Michael Greenfelder/Alamy Images; 15 inset: imageBROKER/Shutterstock; 16–17: Magnus Lundgren/NPL/Minden Pictures; 16 inset: imageBROKER/Shutterstock; 18 top: Hendy Mp/Solent News/Shutterstock; 19 main: Cede Prudente/Getty Images; 20–21: Chien Lee/Minden Pictures; 22 frog: Rushenb/Wikimedia; 23 main: Chien Lee/Minden Pictures; 24–25: Quentin Martinez/Biosphoto/Minden Pictures; 27: Brent Barnes/Stocktrek Images/Getty Images; 28: Ken Griffiths/Getty Images; 29: Michael Leach/Getty Images; 30 top left: Cede Prudente/Getty Images; 30 center: Anthony Pierce/Splashdown/Shutterstock; 30 bottom left: Masahiro Iijima/ardea.com/age fotostock; 30 bottom right: Stephen Dalton/Nature Picture Library/Alamy Images; 32: Scott Linstead/Science Source.

All other photos © Shutterstock.

CONTENTS

INTRODUCTION

ANIMALS THAT FLY

You might have seen birds and insects spread their wings and fly. But did you know that they are not the only animals that can take to the skies? There are a lot of other animals that can fly, too! And some might surprise you.

Animals fly for many different reasons. Flying gets them from place to place. It lets them quickly escape **predators**. It also helps them surprise their **prey**. Some animals fly to snack on food in trees. Get ready to learn about some surprising flyers!

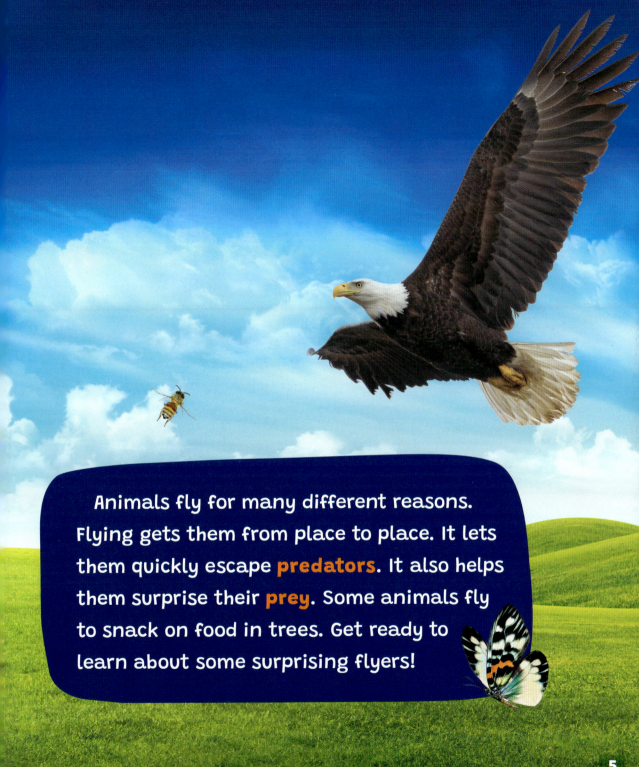

CHAPTER 1

PARADISE TREE SNAKE

FACT FILE

ANIMAL GROUP: Reptile

HABITAT: Wet forests

DIET: Carnivore

HOW BIG?

Paradise tree snakes can grow to be 3 feet (1 m) long.

That is the same length as a baseball bat.

* Ruler not to scale.

FACT!

The paradise tree snake has many names. It is also called the paradise flying snake and the garden flying snake.

This snake can glide for up to 300 feet (91 m).

The paradise tree snake can travel from tree to tree! It travels by **gliding** through the air. How? First it finds a very tall tree. It slithers up to the highest branches. Next, it stretches its body into a flat S shape. Then, it leaps off! It can turn and **steer** its body in the air. This snake glides between trees to catch prey and escape predators.

There are only five snake **species** in the entire world that can fly. The paradise tree snake is one of them. Paradise tree snakes can be found all over South and Southeast Asia. They live in countries such as Thailand and the Philippines.

FACT!

Paradise tree snakes can swallow their prey whole.

These flying snakes eat mostly bats and lizards.

Over time these snakes have gotten used to being around people. Some have been seen flying in public parks and gardens.

CHAPTER 2

DRACO LIZARD

FACT FILE

ANIMAL GROUP: Reptile

HABITAT: Forests

DIET: Insectivore

HOW BIG?

The draco lizard is only 8 inches (20 cm) long.

That is about the same length as a banana.

* Ruler not to scale.

FACT!

Draco lizards can be found in Southeast Asia. They are also known as flying dragons!

The underside of female draco lizards is yellow. Males are blue.

Have you ever seen a flying lizard? The draco lizard can travel up to 30 feet (9 m) through the air at once. That is about half the length of a bowling lane! To fly, the draco lizard unrolls two hidden flaps of skin. They look like wings. It uses these flaps to glide between trees. It uses its long, thin tail to steer.

FACT!

Draco lizards are protective of the trees they live in. They will try to scare other lizards into staying away.

There is a reason why draco lizards fly. They do not like being on the ground! Predators can catch them easily there. Instead, they stay high up in the trees. Draco lizards also fly between trees to eat. Female draco lizards only

leave the trees when it's time to lay eggs. They hurry down to the ground and lay their eggs. They stay with the eggs for one day to protect the nest. Then, the female rushes back up into the trees.

CHAPTER 3

JAPANESE FLYING SQUID

FACT FILE

ANIMAL GROUP: Invertebrate

HABITAT: Ocean

DIET: Carnivore

HOW BIG?

Japanese flying squids are between 12 to 20 inches (30 to 51 cm) long.

That is about the same length as a rolling pin.

* Ruler not to scale.

FACT!

Flying squids have two fins, eight arms, two **tentacles**, and three hearts.

Most Japanese flying squids can be found near the surface of the water.

Did you know some squids can soar out of the ocean? The Japanese flying squid's body has a special **muscle**. It is able to take in water. When the water is pushed back out, it can fly! The force is very powerful. It shoots the squid out of the water! In the air, flying squids can travel more than 30 feet (9 m) per second. They spread out their fins and arms. This helps them glide smoothly.

Flying squids usually fly when being chased by a predator. They can also squirt ink to protect themselves. Their bodies can even change color. Changing color to blend in is called **camouflage**.

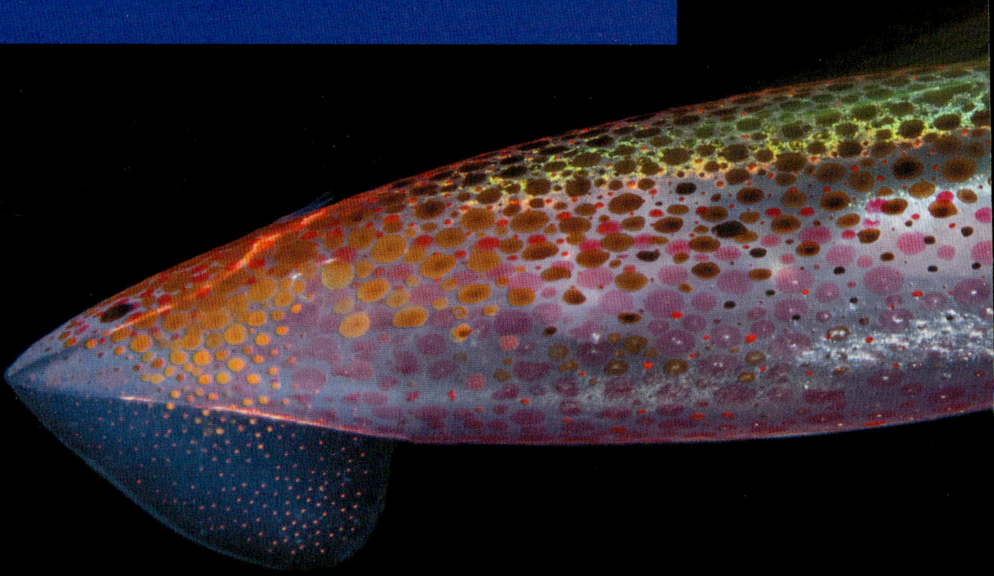

FACT!

Japanese flying squids are most common in the Pacific Ocean.

These squids eat small fish. This helps them survive!

Flying squids use camouflage to hide from predators. They also use it to catch prey.

CHAPTER 4

COLUGO

FACT FILE

ANIMAL GROUP: Mammal

HABITAT: Rainforests

DIET: Herbivore

HOW BIG?

Colugos are 22 to 26 inches (56 to 66 cm) long.

* Ruler not to scale.

This colugo is gliding through a Malaysian forest in Asia.

FACT!

Colugos are **nocturnal**. This means they sleep during the day and are active at night.

The colugo is a skilled glider! To fly, it first uses its sharp claws to climb up a tall tree. When it reaches a high branch, it stretches out its arms and legs as wide as possible. Then it jumps off. The colugo has furry skin that stretches out like a big sheet. This skin helps it glide through the air up to 200 feet (61 m)! That is about the length of a football field.

FACT!

Colugos eat fruits, shoots, and leaves.

This colugo mother holds her baby while she moves.

Why do colugos fly? Because it's faster than running and jumping! It also helps them find food in trees. Some people call colugos flying lemurs. But they are not a type of lemur. Colugos look more like squirrels.

They live in Southeast Asia. Mother colugos fly and carry their babies on their stomachs. The baby colugos grow up! Then they can glide on their own.

CHAPTER 5

WALLACE'S FLYING FROG

FACT FILE

ANIMAL GROUP: Amphibian

HABITAT: Tropical forests

DIET: Insectivore

HOW BIG?

Wallace's flying frogs are 4 inches (10 cm) long.

That is about the same length as a Popsicle stick.

* Ruler not to scale.

FACT!

Other species of frogs can also fly. But the Wallace's flying frog is one of the biggest.

These frogs have shiny green skin.

Some people call this little creature a parachute frog! When it flies, this frog's webbed feet help it travel from tree to tree. On their toes, Wallace's flying frogs have sticky pads. These toe pads let them land gently after gliding. The pads also allow them to stick to tree trunks, so they do not fall.

Wallace's flying frogs love to stay in trees. It's a safe place for them! But tree-climbing snakes like to hunt them. These frogs must be able to fly away fast when they sense danger. They only head to the ground when it's time to find a **mate** and lay eggs.

FACT!

The Wallace's flying frog can glide up to almost 50 feet (15 m).

This flying frog was named after the scientist who discovered it, Alfred R. Wallace.

Wallace's flying frogs can be found in the tropical forests of Malaysia and Borneo in Southeast Asia.

MORE UNEXPECTED FLYERS

MOBULA RAY

Mobula rays are also called devil rays. These fish live in warm oceans. They can leap 6 feet (2 m) in the air. That is about the same height as a refrigerator!

SAILFIN FLYINGFISH

Sailfin flyingfish glide by spreading out their fins. They can fly up to 4 feet (1.2 m) in the air. That is the same length as four rulers! These fish are found in the Atlantic, Pacific, and Indian Oceans.

FLYING FOX

Flying foxes are not actually foxes—they are bats! But they are a special type of bat. They are called megabats. Megabats are a large bat. The largest type of flying fox has a wingspan of up to 6 feet (2 m) in length.

KUHL'S PARACHUTE GECKO

Kuhl's parachute geckos can glide from tree branch to tree branch. They use their flat tails to stay in the air. They can camouflage their skin to make it look like tree bark. This keeps them safe from predators.

KEEP FLYING!

Now you know all about some amazing flying creatures! You learned how snakes slither through the air. You read about how lizards, frogs, and colugos glide on the wind. And you found out how squids use water to fly! Who knew so many animals could take to the skies? You can discover even more creatures that fly by reading about animals. Or, the next time you see an unfamiliar creature, watch it carefully. Maybe it can fly, too!

GLOSSARY

camouflage (KAM-uh-flahzh) a disguise or natural coloring that allows animals to hide by making them look like their surroundings

carnivore (KAHR-nuh-vor) an animal that eats meat

glide to move smoothly and without effort

habitat (HAB-i-tat) the place where an animal or a plant is usually found

herbivore (HUR-buh-vor) an animal that only eats plants

insectivore (in-SEK-tuh-vor) an animal that only eats insects

mate the male or female partner of a pair of animals

muscle (MUHS-uhl) a type of tissue in the body that can contract to produce movement

nocturnal (nahk-TUR-nuhl) active at night

predator (PRED-uh-tur) an animal that lives by hunting other animals for food

prey (pray) an animal that is hunted by another animal for food

species (SPEE-sheez) one of the groups into which plants and animals are divided

steer to make something go in a particular direction

tentacle (TEN-tuh-kuhl) one of the long, flexible limbs of some animals

INDEX

ABOUT THE AUTHOR

Jay Leslie has always loved to write. Everywhere she goes, she carries a notebook just in case she gets a new idea. Most of all, she loves to write the books that she wishes she'd had as a child.